Over the last decade, a series of dramatic research findings have served to emphasize the vital role of the peroxisome. These findings have included major advances in our understanding of its metabolic roles, the genetic specification of peroxisomal components, the complex processes of protein targeting and peroxisomal biogenesis, and the broad involvements of peroxisomal diseases.

This book, by its format, provides both a basic introduction to the peroxisome and its relationship to other components of eukaryotic cells, and a detailed and comprehensive discussion of recent advances. In addition to the topics referred to above, current understanding of the morphology, phylogeny, enzymology and ontogeny of the peroxisome are also reviewed, as are its extraordinary proliferation induced by a variety of drugs and xenobiotics, and its role in hepatocarcinogenesis.

The authors are well respected researchers in this field and have produced an authoritative and readable text, with numerous illustrations and chapter summaries. In this form the text will appeal to advanced undergraduates, postgraduates and researchers in biochemistry, cell biology and biomedical sciences who wish to keep abreast of the many important biological correlations of this organelle.

T0353592

THE PEROXISOME: A VITAL ORGANELLE

THE PEROXISOME:
A VITAL ORGANELLE

COLIN MASTERS AND DENIS CRANE
Faculty of Science and Technology Griffith University, Brisbane

CAMBRIDGE
UNIVERSITY PRESS

CAMBRIDGE UNIVERSITY PRESS
Cambridge, New York, Melbourne, Madrid, Cape Town, Singapore, São Paulo

Cambridge University Press
The Edinburgh Building, Cambridge CB2 8RU, UK

Published in the United States of America by Cambridge University Press, New York

www.cambridge.org
Information on this title: www.cambridge.org/9780521482127

First published 1995
This digitally printed version 2007

A catalogue record for this publication is available from the British Library

Library of Congress Cataloguing in Publication data
Masters, Colin J.
The peroxisome : a vital organelle / Colin Masters and
Denis Crane.
p.   cm.
Includes bibliographical references and index.
ISBN 0 521 48212 7
1. Peroxisomes.   I. Crane, Denis.   II. Title.
QH603.P47M37   1995
574.87′4–dc20   94-40110 CIP

ISBN 978-0-521-48212-7 hardback
ISBN 978-0-521-03683-2 paperback

To Christian de Duve, whose pioneering investigations established the presence and broad significance of the peroxisome, and served as motivation for much of the current research activity in this field.

Contents

Preface

Over the last decade, the peroxisome has moved from the wings of the cellular biology stage to a far more central position – one of evident involvement with many other areas of topical biological interest. A series of dramatic findings have served to establish the essential role of this organelle in the cellular and organismal context, and elevated its ranking in the day-to-day considerations of biomedical scientists.

These findings have included major expansions in the understanding of the metabolic role of this organelle, advances in the genetic analyses of peroxisomal components, insights into the complex processes involved in peroxisomal biogenesis and proliferation, and advances in the broad inter-relationships of peroxisomal disease.

While it is largely through this last aspect (and the film *Lorenzo's Oil*, in particular) that the peroxisome has impacted on the consciousness of the general public, all of these aspects have fostered increasing mutual interest and interaction with other areas of cell biology, and stimulated a broadening interest in the structure and function of the peroxisome.

Despite all these significant advances, though, and the incorporation of peroxisomal involvements into many other diverse areas of topical biological study, an up-to-date, comprehensive reference text has not been available. This book is aimed at meeting this deficiency and providing a contemporary compilation of discoveries and achievements in this field. It includes a brief description of the historical background and perspectives, followed by full accounts of the enzymology of this organelle, the intraparticulate organization of these enzymes, and their genetic and ontogenic analyses. Following on from this, a comprehensive coverage of peroxisomal metabolism has been included, detailing the many roles in the catabolism and anabolism of lipid, carbohydrate and nitrogenous compounds, as well as the extensive involvements in metabolic control. There are also chapters on peroxisomal biogenesis, peroxisomal proliferation and peroxisomal diseases – all areas of widespread, active interest at the moment.

In compiling this treatment, the authors have been conscious of the rapid expansion and broadening of the peroxisomal audience, and the necessity for comfortable access to the relevant peroxisomal information by readers from many and varied backgrounds. For this reason the more detailed discussions in this book have been supplemented by numerous diagrams and illustrations, fully titled references, and summaries and suggestions for further reading at the end of each chapter. These suggestions for further reading have been designed not only to include recent reviews, but also to provide a chronological commentary of major advances in the chapter topics.

In relation to the many experimental procedures that have been described in the literature for the separation of peroxisomal fractions and their biochemical or cytochemical characterization, it may be noted that we have adopted the approach in this text of a general, *in principio* treatment, accompanied by liberal references to the literature on specific methodologies. It was considered that this approach was more appropriate to a general text of this type, rather than the inclusion of a series of lengthy step-by-step descriptions of individual procedures.

It is hoped that in this form the text will be of interest not only to the peroxisomal specialist and to postgraduates in the biomedical sciences who wish to keep abreast of the many significant advances in this important area of cell biology, but also as an optimal item in the reference curriculum of most undergraduate students in biology and medicine.

As another comment on format, it may also be noted that, while many texts in the past have commonly used bidirectional arrows to emphasize the reversible nature of enzyme-catalysed reactions, we have on occasion used single directional arrows in this text. This has been done, not only for purposes of clarity and simplicity, but also in recognition of the trend towards such usage in modern texts.

When a field advances as rapidly as peroxisomal biology, of course, it presents the authors not only with the problem of providing the data in readily accessible form, but also a question (to pursue the initial analogy) of when to ring down the curtain on coverage. In the present instance, it was decided that March 1994 was an appropriate time to terminate our literature references, and readers seeking information published after this date should note that they will need to refer to the original literature.

No book of this nature can be developed without significant interactions with other experts in the field, and we would like to take this opportunity of expressing our gratitude to the many colleagues who have assisted in this undertaking with discussions and encouragement. Several illustrations in the text have originated from the work of fellow researchers, and we have endeavoured to ensure that all these sources have been acknowledged appropriately.

Finally, we wish to thank our respective wives and families, without whose patience, understanding and forbearance, this book would never have been completed.

Colin Masters
Denis Crane

1

Introduction

Outline of the metabolic role

With the recognition of the indispensable nature of the peroxisomal contribution to many metabolic processes in plants and animals, and the establishment of the broad clinical significance of the peroxisomal disorders in humans, the vital role of these organelles in multicellular organisms has attained wide recognition. The peroxisome has become the focus of much contemporary research interest, has attracted an increased topical consideration in relation to the other major structures in eukaryotic cells, and become an essential component of the background knowledge and curricula of many cell and medical biologists. It is with this emerging role in mind that this introduction sketches an outline of the peroxisomal involvement in cell biology, leaving a more detailed treatment of individual aspects of structure and function to subsequent chapters.

Overall, then, peroxisomes may be said to be widely distributed in both the plant and animal kingdoms, and to make an important contribution to a variety of catabolic and anabolic processes. These organelles were originally characterized by their content of catalase and several oxidative enzymes (de Duve & Baudhuin, 1966; Masters & Holmes, 1977; Böck *et al.*, 1980; Masters & Crane, 1992a,b) and, as the name was intended to suggest, the functions of this particle were envisaged as encompassing the generation and decomposition of hydrogen peroxide. Because of structural similarities and this broad biochemical specification, an overlap of terminology was possible between such commonly applied descriptions of subcellular structures as microbodies, peroxisomes and glyoxysomes. As de Duve (1969) has made clear, however, the term microbody was intended to describe only the morphological appearance, and almost certainly includes other biochemically unrelated particles, while the term glyoxysome is restrictive in that it refers to a special class of peroxisome that stresses the metabolism of glyoxylate. Smaller forms of peroxisomes (microperoxisomes) have also been reported, but current usage tends to

1

employ the term peroxisome to embrace both glyoxysomes and microperoxisomes.

The functions of this biochemically diverse organelle vary from one organism to another. Peroxisomes carry out a variety of metabolic functions that vary with cell type and environmental conditions. In animal cells, these functions include respiration based on the hydrogen peroxide-forming oxidases and catalase, fatty acid oxidation, plasmalogen biosynthesis, alcohol oxidation, transaminations, and the metabolism of purines, polyamines, bile acids and other substrates.

In addition to these important metabolic roles in mammalian tissues, it should be noted that further significant roles occur in both the catabolic and anabolic pathways of plants (Tolbert, 1971). These additional roles are possible because of the distinctive enzyme compositions in these different life forms. In seeds rich in lipids, peroxisomes (glyoxysomes) are the site of the breakdown of fatty acids to succinate via the glyoxylate cycle, and participate in gluconeogenesis in this way. In leaf tissues, peroxisomes serve as sites of photorespiration as well.

Biochemical definition

The identification and biochemical definition of the peroxisome arose from complementary cytological and biochemical investigations. In 1954 Rhodin described the existence of cytoplasmic organelles in renal tubules of the mouse; they were surrounded by a single limiting membrane and contained a finely granular matrix. He named these organelles 'microbodies', and, subsequently, similar cytoplasmic inclusions were described in hepatic and other cell types (Fig. 1.1).

About the same time, biochemical studies demonstrated the existence of a particle in mammalian liver that contained catalase and several oxidases. After an initial description of the particulate nature of urate oxidase, Novikoff and Goldfischer (1969) demonstrated that the centrifugal behaviour of this enzyme was distinct from that of mitochondria, but resembled that of the lysosomal enzyme acid phosphatase. Similar sedimentation properties were subsequently described for D-amino acid oxidase and catalase. That catalase and these oxidases were localized in a previously undescribed particle, distinct from lysosomes, was demonstrated in an elegant series of experiments by de Duve and coworkers, involving density gradient and latency studies (de Duve and Baudhuin, 1966), and in the same period a method of altering the sedimentation properties of lysosomes by treatment with a non-ionic detergent, Triton WR-1339, was reported

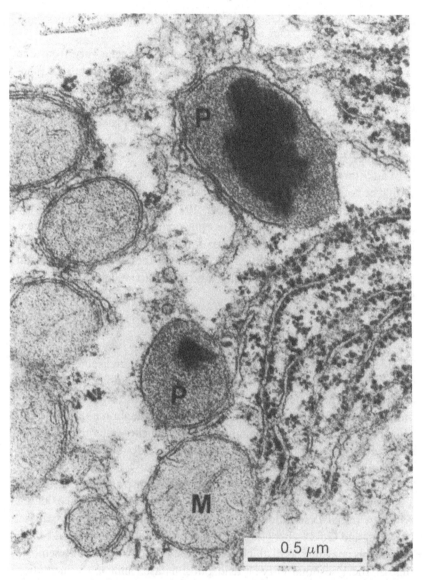

Fig. 1.1. Electron micrograph of peroxisomes in rat liver. Note the single limiting membrane, the electron-dense crystalline core, and the fine granular matrix for each of the peroxisomes (P) shown. A mitochondrion (M) and membranes of the endoplasmic reticulum with associated ribosomes are also shown in the bottom right of the picture. (Contributed by J. Hughes, Adelaide Children's Hospital.)

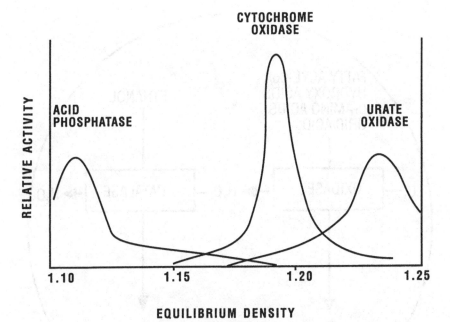

Fig. 1.2. An illustration of the separation of peroxisomes from lysosomes and mitochondria by centrifugation of rat liver homogenate in a sucrose gradient. Rats were injected with Triton WR-1339 – a treatment that altered the density of the lysosomes (marker enzyme, acid phosphatase), so that they could be clearly separated from mitochondria (marker enzyme, cytochrome oxidase) and peroxisomes (marker enzyme, urate oxidase). Generally, density equilibration of normal liver homogenates leads to substantial overlapping of these organelle fractions in the density range of 1.20–1.25.

(Wattiaux *et al.*, 1963). This procedure allowed the successful separation of peroxisomes from both lysosomes and mitochondria in a sucrose density gradient, and their identification with the previously described microbodies (Fig. 1.2). Further confirmation was provided by the development of cytochemical staining techniques for these enzymes (Novikoff and Goldfischer, 1969), and it subsequently became evident that the peroxisome was widely distributed within both animal and plant kingdoms.

After the successful demonstration of the occurrence of this separate organelle, considerable research effort was directed toward elucidation of the details of its function. Early investigators were impressed by the fact that the oxidases in these organelles produce hydrogen peroxide as a result of their enzymic activities, and that this hydrogen peroxide may be reduced to water by catalase, which is also present in these organelles. The electron donors in the latter process may be substances such as methanol, ethanol,

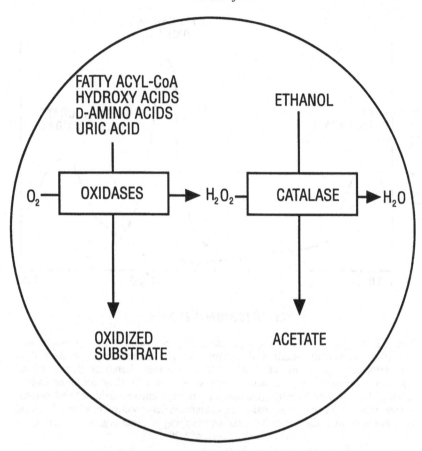

Fig. 1.3. An illustration of the concept of peroxisomal metabolism. Substrates of peroxisomal oxidases are metabolized, with the production of hydrogen peroxide, which in turn may be catalytically decomposed with or without the attendant oxidation of substrates such as ethanol.

nitrite, formate or hydrogen peroxide itself. Hence it seemed that a form of respiration was possible in which electrons removed from these various metabolites are added to oxygen, resulting in the formation of water. It was on the basis of such considerations that de Duve (1969) proposed that these organelles be termed peroxisomes, and this concept has proved to be of great value in providing a linkage between the variety of microbody types that has been found to be present throughout the eukaryotes (Fig. 1.3).

Quite independently of these peroxisomal definitions in animal tissues, investigations of the glyoxylate cycle in germinating castor beans by Briedenbach and Beevers (1967) led to the discovery of another subcellular

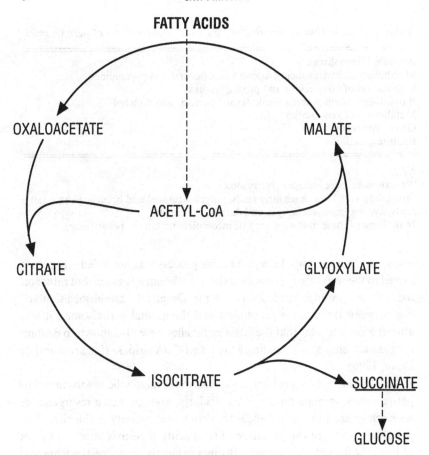

FATTY ACIDS

OXALOACETATE   MALATE

ACETYL-CoA

CITRATE   GLYOXYLATE

ISOCITRATE   SUCCINATE

GLUCOSE

Fig. 1.4. The operation of the glyoxylate cycle in the conversion of fatty acids to glucose in plants.

particle with the general characteristics of a microbody. These particles contained the enzymes of the glyoxylate cycle, and hence were given the name glyoxysomes. Subsequently, similarities between the biochemical composition of plant glyoxysomes and animal peroxisomes in other respects were noted, and catalase and some oxidases were identified in glyoxysomes. Thus glyoxysomes were established as a form of peroxisome, although the separate terminologies are quite often retained to indicate the presence or absence of glyoxylate cycle enzymes in this organelle (Fig. 1.4).

Following on from these early studies on the glyoxysome, Cooper and Beevers (1969) made the further discovery that glyoxysomes in germinating seedlings also have the enzymic capacity to cause the β-oxidation of fatty

Table 1.1. *An outline of the diverse metabolic involvements of peroxisomes*

Peroxide metabolism
Metabolism of hydroxyacids, amino acids, purines and pyrimidines
β-Oxidation of fatty acids *ᵃ* and prostaglandins
Biosynthesis of ether lipids, cholesterol, bile acids and dolichol
Metabolism of xenobiotics
Glyoxylate metabolism
Photorespiration

Notes:
'Peroxisomes' here includes glyoxysomes.
ᵃ Including very long chain fatty acids, polyunsaturated and branched chain fatty acids, hydroxymonocarboxylic and dicarboxylic acids.
N.B. Some of these pathways may be incomplete within the peroxisome.

acids, but that the initial enzyme in this process was not a dehydrogenase linked to the respiratory chain as in the mitochondrial system, but rather an oxidase that produced hydrogen peroxide. Despite the established similarities between the plant glyoxysomes and the animal peroxisomes, it was almost a decade later that the latter organelles were also shown to contain β-oxidation enzymes, including a fatty acyl-CoA oxidase (Lazarow and de Duve, 1976).

This finding led to a realization of the wider metabolic involvements of peroxisomes, in higher animals particularly, and sparked a resurgence of research interest in this organelle that continues actively to this time. The metabolic route for the oxidation of fatty acids in peroxisomes was shown to be a true β-oxidation system – distinct in metabolic properties from that previously demonstrated in mitochondria, yet analogous in many aspects, and the contribution of peroxisomal oxidation to a diversity of metabolic process was subsequently established. These processes included the oxidation of very long chain fatty acids and xenobiotics, the cleavage of the cholesterol side-chain in the course of bile acid formation, the shortening of the carbon chain of long chain dicarboxylic and ω-hydroxymonocarboxylic acids, the catabolism of prostaglandins, and a role in anabolic functions such as the synthesis of cholesterol and ether glycerolipids (Masters and Crane, 1992a)(see Chapters 5 and 6; Table 1.1).

In addition to metabolism, the enzymological characteristics of these organelles, the internal disposition of constituent proteins, the properties of the membrane, peroxisomal proliferation and biogenesis, and the broad involvement of these organelles in disease processes are all attracting considerable attention. Inherited disorders affecting human peroxisomes

were first demonstrated by Goldfischer *et al.* (1973) in the case of Zell-
weger's cerebrohepatorenal syndrome. In this condition, recognizable
peroxisomes are absent from the patient's tissues, and the prognosis is
death in early infancy. A number of peroxisomal enzyme deficiencies have
since been shown to correspond to human genetic disorders, and the clinical
ramifications of peroxisomal deficiencies have expanded to hold a broad
biomedical significance. Current knowledge of these aspects of peroxisomal
biology are described in the subsequent chapters of this book, but in order
to comprehend some of the difficulties involved in a definition of the
physiological role of the peroxisome it is first necessary to consider some of
the morphological and phylogenetic ramifications.

Morphology

The first description of the peroxisome (Rhodin, 1954; Fig. 1.5) listed these
organelles as:

spheric or oval bodies between the mitochondria. The number, size and shape of the
microbodies varies from cell to cell with a mean of about ten in each cell. The mean
length is 0.3 mμ and the mean width 0.1 mμ. They are surrounded by an osmium
impregnated single membrane, 45 Å thick. The ground substance of the microbodies exhibits the same opacity as the stroma in the mitochondria and consists of a
finely granular structure, the size of the granules being about 40–50 Å.

Succeeding studies of peroxisomes in liver revealed the presence of an
electron-dense central core, or crystalloid, which, because of its common
occurrence in peroxisomes of mammalian hepatocytes in non-humanoid
species, came to be quoted as the most distinctive structural feature of these
organelles. The core was later shown to be a complex polytubular structure
(Hruban and Swift, 1964). An additional structure, the marginal plate, was
also reported in some cell types (Schnitka, 1966), and in methylotropic
yeasts, too, growth on methanol-containing nutrient induces striking cubic
crystals of alcohol oxidase.

   Peroxisomes are commonly described in the literature as being spherical
or ovoid, and may exhibit a wide range of sizes (from 0.1 to 1.5 μm in
diameter). The normal size range reported for liver and kidney is between
0.3 and 0.9 μm, and similar sizes have been reported for other tissues. Most
hepatocytes contain between 400 and 600 peroxisomes, and these organelles occupy about 2% of the cell volume.

   The limiting membrane of peroxisomes is a single membrane, consisting
of a triple-layered structure of thickness 4.5 to 8 nm. This is thinner than
that of lysosomes, the plasma membrane and most other single membrane

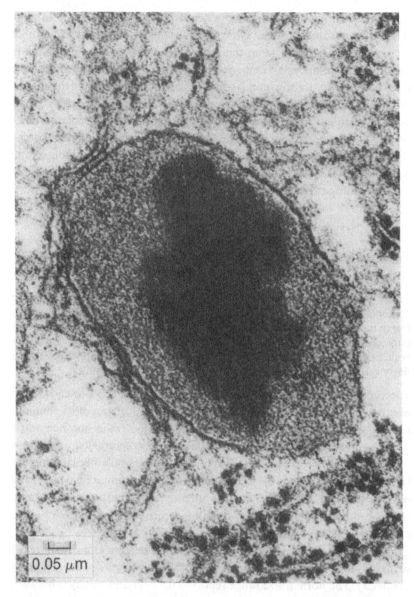

Fig. 1.5. Magnification of the electron micrograph of one of the rat liver peroxisomes from Fig. 1.1 to highlight the morphological features of these organelles.

bound cell inclusions, but is of comparable thickness to the endoplasmic reticulum (Beard and Allen, 1968). An important differential property of the peroxisomal membrane is that it exhibits quite different permeability properties *in vitro* from those of other subcellular organelles. The membrane of isolated peroxisomes is highly permeable to small molecules such as sucrose, small substrates and inorganic ions; moreover, it is more resistant to lysis by digitonin, and susceptible to disruption by physical treatments such as freezing, high speed laminar shearing or sonic oscillation (Masters and Holmes, 1977).

The occurrence of nucleoids in peroxisomes is in most cases related to the occurrence of urate oxidase. Biochemical studies have shown, for example, that this insoluble enzyme is associated with the crystalline core in liver and kidney, while species have been reported that lack this enzyme and are also devoid of nucleoids. There are noteworthy exceptions, however, to this concomitant occurrence of urate oxidase and nucleoids. Rat kidney and human kidney, for example, exhibit no urate oxidase activity, but do exhibit nucleoid structures, and some non-mammalian vertebrates have high levels of hepatic urate oxidase but do not display correspondingly high numbers of crystalline cores (Masters and Holmes, 1977). There are also reports of other enzymes, such as xanthine oxidase, occurring in some peroxisomal cores (Angermüller *et al.*, 1987), and of urate oxidase activity occurring in the peroxisomal matrix as well as in the core (Van den Munckhof *et al.*, 1994).

In some species, as mentioned, a marginal plate has been described in peroxisomes. This plate resembles a flat, uniformly thick structure, located at the periphery of the peroxisome, and separated from the inner surface of the membrane by a narrow space of lower density than the surrounding matrix. Recently it has been demonstrated that L-α-hydroxyacid oxidase B is localized in this structure (Zaar *et al.*, 1991).

Several early morphological studies also reported that peroxisomes may exhibit direct connections with the endoplasmic reticulum – continuities and membrane protrusions often being referred to (Novikoff and Shin, 1964) – and such observations led to a view that these channels may play an important role in the delivery of peroxisomal proteins from the endoplasmic reticulum during biogenesis, as described in chapter 8 (see also Fig. 1.6). There are, however, many difficulties in distinguishing between peroxisomal membranes and the endoplasmic reticulum on morphological grounds alone, and these apparent continuities have come increasingly into question more recently. Shio and Lazarow (1981), for example, have used cytochemical stains in these areas of apparent continuity to test for specific

Index

Italic numbers indicate pages with figures.